Copyright Information

A Day in the Life of a Scrum Master

OctaneArmyPublishing@gmail.com

www.agilearmy.com

https://www.facebook.com/agilearmy/

Table of Contents

Copyright Information... 1
Foreword.. 3
A Day (week) in the Life of a Scrum Master: 7
10 Traits of an effective Scrum Master:.................... 24
Tools that I use every day: 35
The path to becoming a Scrum Master: 37
FAQs.. 40
Final Thoughts 44

Foreword

Welcome and congratulations on taking the first steps into becoming a Scrum Master! If you are reading this, you are most likely interested in the career, curious about what other Scrum Masters do on a daily basis, or perhaps you just thought it was a funny title, and you wanted to read a little bit about it. Whatever the reason happens to be, it is my goal to give you an accurate portrayal and precise account of what a Scrum Master does in a typical sprint. Unlike other scrum publications, I won't focus in depth on the ceremonies and the structure of how everything should go; Instead, I want to give you real life examples of how I go about my daily schedule working with my teams and helping to improve my department with the help of the Scrum framework. By the end of this book, I hope that I have given you enough insight into whether this role sounds exciting and worthwhile for you to pursue. Without any more hesitation, let's jump right into the world of Scrum!

"You're a what???" is usually the most common response I receive when I tell someone about my profession. Scrum Master is a funny title for a serious role. In the world of agile project development, a Scrum Master is an important pillar into developing strong teams and ensuring that projects get delivered on time, under budget and to the customer's specification. Although many may think that the role is a management position, it is not; In fact, it is a *leadership* position; one that can take a team to new heights or be the reason why a project fails. Throughout my training, I was constantly taught about how a Scrum Master is a servant leader to a team, and ***not*** a dictator. One of the most important principles that is preached and reiterated in scrum is about empowering self-organizing teams to make decisions about what they work on, how they divide up the work, and how much work they can take on at a given moment.

If you are brand new to Scrum, it is then necessary to understand the components of a typical development team. In the organization that I work for, we deal with very technical and complex software. The teams I serve are generally made up of a Scrum Master, a Product Owner, and a development team that usually consists of various skill sets and knowledge of different programming and scripting languages. The team will also include a QA engineer to ensure that all the development work is checked for errors and bugs that could hinder the release of future releases. The Product Owner is responsible for bringing future work to the team, explaining the business value, and then working with the different team members to break down the work into sizable chunks in order to pave the way for an efficient sprint. It is a goal of the Scrum Master to help organize all the different entities of the team to ensure that communication, coordination and synergy occurs. With the right amount of leadership, a strong team will soar to new heights. When a team feels empowered and given space to tackle their work, they will feel motivated to succeed and prosper.

A Day (week) in the Life of a Scrum Master:

The following is an accurate representation of what my typical day/week looks like from the perspective of a Scrum Master. Names, titles, and project details have been altered to preserve the anonymity of my fellow coworkers and to avoid sharing any sensitive company data. Rather than highlight every single repetitive meeting, I sampled some of the most important meetings from the week.

My Background:

I currently work as a Scrum Master for a well-established software company. My department is large enough that we have around 10 development teams and work in a Scaled Agile Framework (SAFe). Each of our development teams consists of Scrum Masters, Product Owners, UI & back-end developers and QA engineers.

Monday: 10am, Bug Triage Meeting

During the early months of my tenure at the software company I work for, we were struggling to handle the incoming flux of new bugs and defects. Bugs and defects are found by everyone from developers, SE's, customers, and anyone else who uses our product on a regular basis. In order to improve our visibility on the critical bugs that recently emerged, we instituted this triage meeting every morning in order to discuss and prioritize the latest most pressing bugs and defects. We have members from every squad attend in order to provide their own insight and offer advice on who should take on the work and possibly how to resolve the issues. I attend this meeting every morning so that I have prior knowledge of when new work will be impacting my team. Depending on our current backlog, I will discuss the possible work with our

Product Owner and determine the best course of action so that our team does not take on work that might negatively affect their current sprint work.

During this particular meeting, a critical bug was reported by one of our biggest clients. There seemed to be a memory issue with one of their appliances. For unexpected reasons, their RAM began thrashing and would spike to the point that their specific software release became unusable. After doing some initial discussion, the bug was assigned to the platform team because of the possibility that a recent change could have caused something unexpected to occur. Due the criticality of the bug and the clout of the customer, we recognized the urgency to take on the bug so that we could begin working on it.

Monday: 10:45am, Platform Team Stand-up Meeting

After attending the bug triage meeting, I knew there was going to be a lot of discussion concerning the new bug that was just assigned to the team. We went through our typical routine and had each team member update the rest of the team on their status, which involves, what they worked on yesterday, what they are working on today, and whether they have any blockers that may be hindering their work. We have a working agreement that we will "timebox" these meetings to 15 minutes so that we don't get stuck in the weeds with details on specific topics. After the allotted 15 minutes, we allowed the extraneous stakeholders attending the meeting to leave while the rest of the team discussed the best plan of action on how to research the bug, who the bug was going to be specifically assigned to, and how they were going to go about solving the problem. In this situation, we had a developer who had finished his sprint work and had the necessary bandwidth to begin working on the bug. Throughout these types of meetings, I make an emphasis to always facilitate the discussion so that the team can gain the most value from what's being talked about, while at the same time, minimizing the amount of time that is taken up so that the developers can get back to their daily work. It is also necessary to make sure that I ask the tough questions and confirm that the bug assignee has the right information and vision in order to tackle the task head on.

Monday: 11:30am, Feature Work Meeting

One of the fantastic disciplines that our department has is always planning for features many weeks and months in advance. We do this so that when we get to the actual feature work, the stories are well defined and all the risks, assumptions, and goals are clear and understood. This specific meeting regarded a new feature that was promised to a huge client of ours. We had been having concerns about the authentication and authorization aspect of this feature. Due to the necessity of maintaining the highest security standards in our product, we wanted to make sure that nothing was being forgotten with the implementation. We brought in a security advocate to our meeting to help answer questions and offer advice on how to bake the proper security measures into the new feature. The byproduct of this meeting was a lot of diagrams on the white board, important notes about the implementation and a strategy on how to attack this portion of the feature.

Monday 1:00pm, Customer Call Meeting

Another one of the standard initiatives that our company promotes is having phone calls and videoconference meetings with our clients so that we have the opportunity to watch them use our product and see where their pain points are. This allows us a rare glimpse to see how our customers are truly using our software. During this particular meeting, we were given access to their beta lab so that we could dig around and figure out why a specific beta build was causing problems. I had coordinated with a professional service representative, our product owner and one of our developers beforehand so that we were all informed on the issue and what we needed to be looking for during the call. The preparation paid off and the investigation went smoothly. We were able to immediately solve the client's problem and found a few issues that could be fixed in future releases. By preemptively going straight to the customer, we are able to save a lot of time and resources and make our customer very happy in the end.

Monday 2:15pm, Architect Team Stand-up Meeting

The other team that I serve focuses on the architecture of our software. They are responsible for creating the vision for future initiatives and helping out other teams when they have questions about how to solve a specific problem. During this stand-up, we talked about the challenges associated with the work they are currently doing to create a new micro-service. This team also has a few remote team members, so I make sure that everyone has been communicating with each other and that the work is staying on track. They are a very proficient team and have many years of experience, but I always find it beneficial for everyone involved to ask a lot of questions so that I understand the scope of their work and also so that the other team members bring up comments about their individual tasks in case they have questions for others.

Tuesday 9:30am, Product Owner/Scrum Master Joint Meeting

Every Tuesday and Thursday morning, we hold a joint Product Owner/Scrum Master meeting so that we can discuss what each of our teams are working on and offer help to teams who may need assistance from others. Working in a complex environment, it is very easy to lose track of what other teams are currently engaged in. On this particular morning, I discussed the challenges of working with one of our offshore development teams that was currently helping the platform team with automation stories. Because of the time difference and distance, it has made it very difficult to ensure that the offshore team understands the work and is comfortable with how to complete it. I asked for advice from fellow Scrum Masters and Product Owners who had experienced similar issues in the past. Overall, it is a great opportunity to help each other through challenges that each team is facing. I left the meeting with a few tips on what our team could do better to keep the communication open between the local and offshore teams and better accomplish the mission at hand.

Tuesday 11:00am, Leadership Meeting

One of the most important meetings that I look forward to every two weeks is our leadership meeting. It offers a chance for my Product Owner and I to sit down with our director and discuss ways to improve our team. We talk about everything from team dynamics, current sprint work, and topics that may affect our team in the future. During this week's leadership meeting, we spoke heavily about to improve our stand-ups to ensure that everyone is engaged and getting something out of the meeting. It is so easy to go through the motions and leave the meeting without actually getting any value from it. Our director offered us advice on how to structure the meeting to elicit better responses from some of our quieter team members. We left the meeting with a few new items to work on and would report back at the next meeting with our progress.

Tuesday 1:00pm, Backlog Grooming

This may be one of the least exciting meetings throughout the week but it is definitely one of the most important meetings. We use the backlog grooming time to break down stories in our backlog and prepare them for the upcoming sprint. It is crucial that the team takes this time seriously so that all the stories have adequate details, acceptance criteria, and are overall well defined. We used this last backlog-grooming meeting of the sprint to finalize a few stories that we would be working on in the next sprint. Although the Product Owner leads this meeting, it is vital that the Scrum Master attends and ensures that the right questions are being asked and that the team members truly understand what is expected from each story.

Tuesday 4:00pm, Platform Team Demo

I always get excited on demo day because it allows the teams to show off their most recent work to other teams and a few vital stakeholders in the company. This sprint's demo was especially exciting because the platform team unveiled a new tool that allows developers to quickly spin up multiple virtual machines with the software version of their choice in a matter of seconds. The process to do this before was always agonizing since it required many different steps and multiple programs to make it happen. The room was packed to see this new tool and all the developers left the demo extremely excited knowing that they would be able to save hours every single sprint because of this brand new tool that was now completed.

Tuesday 4:30pm, Platform Team Retrospective

Immediately following our team's demo, we hold our retrospective meeting to discuss the positives and negatives of the team's completed sprint. As a Scrum Master, I take a lot of pride in facilitating this meeting because I know how important it is to reflect on our performance. I encourage the team to be as honest as possible and find ways to constantly improve. One of the reoccurring points that we discuss every sprint is how to stabilize our velocity and ensure that we are not consistently rolling over stories to the next sprint. It is all too common for development teams to take on more work than they can handle and then find themselves unable to complete all the work for the sprint. During this retrospective, we talked about ways to better size our stories so that we don't run into the same issue in future sprints. At the end of the meeting, I published all of our comments on our team's wiki page so that we can review our thoughts at future meetings.

Wednesday 10:30am, Platform Team Sprint Planning

This particular Wednesday is the beginning of our next sprint. For many, this is an exciting time because it means that they get to work on new stories and tasks. This meeting went smoothly because the Product Owner and team had done a fantastic job grooming the backlog in preparation for this sprint. During this meeting, I take the responsibility of navigating through our team's Rally page and pulling in stories, updating the hours and making a few adjustments to a few tasks. If you are not familiar with Rally, it is the website we use to help track all the work that we do as an organization. Every team that works with our product documents their work in Rally in order to stay organized and provide transparency to the rest of the company. I made sure to make a few adjustments to a few of the team member's capacity in this sprint because they knew they were going to be out on vacation for a day, which would decrease the amount of work they were capable of taking on. With the previous day's retrospective still on our minds, we worked hard to come up with a good velocity for the team and appropriate capacities for each of the team members.

Thursday 1:00pm, Scrum Master Strategic Meeting

During our Strategic meetings, the Scrum Masters get together with our director to discuss topics that are affecting our department. During this meeting, we talked about a very important issue regarding the fragility of the current software build process. Because of various reasons such as network issues, lack of VM space, and unit test failures, our builds will often break at a moment's notice. We took this time to discuss ways that we could improve the current process so that the teams did not have to waste so much time constantly fixing the builds. We came up with a plan to audit the Continuous Integration (CI) build process that the architecture team currently uses with the hopes of gaining some insight on their methods and workflow. We each left the meeting with different tasks that we were responsible for before the next strategic meeting.

Everyday Tasks

When I am not in meetings and participating in ceremonies, I fill in my time with typical work such as catching up on emails, talking with other teams about their work, getting rid of blockers that may be hindering my team and of course socializing and having a good time with my fellow coworkers. At the end of the day, your teams will be so much stronger and more effective if everyone enjoys being around each other. We love to joke around, go out for team lunches, have Nerf gun fights and talk about whatever is currently going on in the company and in the world. I have been fortunate enough to work with teams that are all very humble and hard working and are constantly looking out for each other. As a Scrum Master, this is the type of environment that you want to create so that your team members enjoy coming to work every day.

10 Traits of an effective Scrum Master:

One of the misconceptions about being a Scrum Master is that you can just go through the ceremonies on a daily and weekly basis and expect everything to run smoothly. Being an effective Scrum Master requires many important traits in order to help lead your team(s) to success. From my experience, here are a few traits that are necessary for elevating your role as a Scrum Master.

Leadership:

One of my favorite quotes regarding leadership comes from two amazing American heroes, Jocko Willink and Leif Babin. During their time serving as Navy Seals, they trained others around them about how "there are no bad teams, only bad leaders". When working with your Scrum teams, you may not always have everyone on the same page, or have the same strengths as others. It is imperative that as a great leader, you take charge and find ways to motivate and inspire your teams to work hard and effectively, despite the shortcomings they may have. Great leaders find a way to win even in the toughest situations. If your team fails, it is because those above them did not lead effectively.

Good Listener:

This is often one of the most overlooked traits that is necessary to not only be a great Scrum Master, but to be a great leader in general. All too often, people assume that those who know the most or have the most intelligent things to say are the ones to admire. Instead, focus on being a good listener and really understanding what people are saying to you that you can help them with. At the core of every human interaction, people want to feel appreciated and listened to. If you can show others that you are truly interested in what they have to say, you will gain their trust and they will be more likely to go out of the way to work hard for you and return the favor.

Emotional intelligence:

Emotional intelligence is a buzzword that you have probably often read about in business books and heard about in self-development seminars. The reason why it is mentioned so frequently is because of its importance with being successful. Emotional intelligence is the ability to manage your own feelings and recognize and react to other people's emotions in an appropriate manner. It is not uncommon for people to completely misread situations with others and then feel confounded when there are conflicts and a lack of communication. As a leader, it is imperative that you have the ability to work with people of all backgrounds, skill levels, education and age. Having strong emotional intelligence will give you the opportunity to succeed in tough environments and maximize the teams you work with. You can improve your emotional intelligence by taking the time to truly listen to your co-workers, find out what they do well, what they struggle with, and what motivates them. Be a good listener and show interest in what they have to say. When issues or problems arise, take the necessary time to *Stop*, *Think*, *Observe* and *Plan*. By reflecting on problems in this manner, it will increase your chances in responding effectively and allow you to avoid making any rash decisions. As a Scrum Master, it is essential that you build a strong bond with members of your team and learn what makes him or her tick so that you can help maximize their value to the team and to your organization.

Conflict Resolution:

No matter what industry you work in, there will always be conflicts that occur between co-workers. As a Scrum Master, you are the first line of defense when there is in-fighting on your team(s). It is not uncommon for teammates to have disagreements with how to approach certain projects or how long a certain task may take to be completed. Sometimes, team members will feel inferior or annoyed by another member because of their work ethic or personality. An experienced Scrum Master will recognize these problems before the conflicts get out of hand and find a way to mitigate the situation. This may require individual conversations with the members in question and even a private meeting between the parties in order to find the best way to come to a resolution. It is often suggested that these sensitive subjects should be handled out of the visibility of the rest of the team in order to allow for trust and privacy to remain intact. If the problem affects the whole team, it is then best to include the team, but still in a private and responsible manner.

Stand up for their team:

Every team wants to know that their Scrum Master has their back. When management comes down with new deadlines or unrealistic expectations, the Scrum Master has to know when to step in and voice the team's concerns. After all, one of the Scrum Master's main responsibilities is to shield the team from outside factors. The Scrum Master should know his team better than any other person in the organization and therefore know what they are capable of, when they need assistance, and what is not feasible. When a team knows that they have the support of their Scrum Master, they will trust him/her more, which creates a stronger more unified team.

Know when to say "No":

One of the hardest things to say to your superiors or teammates is "No". We all want to please others and be a good team player, but there comes a time when it is imperative that you are able to say "No" when issues arise that could adversely hurt your team. Similar to the idea of standing up for your team, an effective Scrum Master has the foresight to tell product management "No" when they try to slide more work on the team, or when management wants to make an important decision regarding deadlines without consulting the rest of the department. Sometimes you may have to give tough love to a team member and strike down a request that could affect the rest of the team. It may cause temporary pain but in the long run, being able to make the tough calls will ultimately improve your team's performance.

Organized:

Companies employ the use of scrum because they most likely have complex projects that need to be completed. With the incredible amount of moving parts and rapidly changing environments, a Scrum Master must be organized in order to keep up with the constant changes and tasks that need to be taken care of. This starts with properly taking notes at your meetings, responding to emails in a timely fashion, keeping a list of tasks that need to be taken care of and keeping your daily calendar up to date. If you don't have to worry about losing track of important notes, dates and tasks, you can more effectively focus on helping your team.

Good Communicator:

As a Scrum Master, you will be in charge of facilitating many ceremonies, which will require you to lead discussions and direct conversations in the right direction. You will also most likely work with some developers who may not have the best communication skills. It is crucial that you feel comfortable speaking in front of people and that you can get your point across in a way that is understandable to the audience and effective with its message. It is not uncommon to have to break down a very complex topic into a simple and to-the-point statement. This is not always an easy task to do, but the best Scrum Masters recognize the necessity to always be improving their communication skills and continually learning in order to be more aware of their subject matter so that they can feel confident when they need to speak about the projects their teams are working on.

Confidence:

One of the most challenging parts of being a Scrum Master is being able to have meaningful and serious conversations with others about topics that you may not have much knowledge about. For example, in the software industry, you may be assigned to lead a team of well-established software architects that have 20+ years of experience. You will need confidence to be able to question them when you feel they are going off track or propose ideas that are not in line with the goals of the project. It is imperative that you feel comfortable in challenging those who have many more years experience than you. By always striving to learn more and asking the right questions, you will gain the trust of your teams and improve your confidence so that the tough conversations will be a lot easier to handle.

Flexibility:

Every work environment will have its own challenges and dynamics. Team members will have their own unique issues as well. Despite all the planning and preparation you account for, there will ultimately be changes, setbacks, and unforeseen circumstances. As an effective Scrum Master, you must remain flexible and be able to go with the flow on a daily basis. Working in an agile environment means that you will have constant unforeseen changes. Flexibility will be a necessary trait to have in order to thrive in a complex business environment.

Tools that I use every day:

Although software tools will vary from organization to organization, a few of the programs that I use on a daily basis are the following:

Rally (Now named CA Technologies): Enterprise-class software and services solutions to drive business agility. We use it to track all of our organization's work and it provides visibility to everyone who helps create our software.
https://www.rallydev.com/

JIRA: Jira is a proprietary issue-tracking product, developed by Atlassian. It provides bug tracking, issue tracking, and project management functions.
https://www.atlassian.com/software/jira

Confluence: Confluence is our company's internal wiki that we use to post development notes, blogs,

schedules and interesting articles regarding our software.
https://www.atlassian.com/software/confluence

Slack: This is our internal chat messaging system that we use to quickly share ideas and comments with other team members. It helps improve communication and it cuts down on the number of emails that must be sent.
https://slack.com/

WebEx: WebEx provides on-demand collaboration, online meeting, web conferencing and videoconferencing applications.
https://www.webex.com/

The path to becoming a Scrum Master:

Landing a scrum master position is not the easiest role to obtain. Organizations are often looking for individuals with project management and leadership experience. To make matters more complicated, few organizations have adapted an agile framework, which makes Scrum Masters a rare breed. If you do not have the project management or leadership experience but you have decided that a Scrum Master is a position that you would like to apply for, the first step to improve your resume should be to become a "Certified Scrum Master" or a CSM. During the certification course, you will spend a weekend going through the mechanics and learning what a Scrum Master does and how to effectively lead an agile team. At the end of the weekend, you will need to pass an exam to complete your certification. Once you have your certification, you will be better prepared for your interviews and stand a better chance of landing the role since you have been legitimized by the scrum community.

Another less traveled route to become a Scrum Master is to intern underneath a current Scrum Master or agile coach and learn how they operate and lead their teams. This is a great way to gain experience without having the pressure associated with being a full-time Scrum Master. I was lucky enough to have the opportunity to shadow a fantastic agile coach/Scrum Master during my internship. Within a few months, I felt comfortable taking on my own teams and spreading my wings. As a result of my internship, I was offered a full-time role and it has allowed me the opportunity to improve my leadership skills and pave the way for many different career paths going forward.

The job market for Scrum Masters is hot right now. According to Glassdoor, the national average salary for a Scrum Master is $91,440 with many experienced scrum professionals earning salaries in the low to mid six-figure range. If you are interested in becoming a Scrum Master and want to learn more about the certification process, check out https://www.scrumalliance.org/.

There are many other great resources that can improve your skills as a Scrum Master. If you want to read a book by one of the "fathers of the scrum movement", check out "Scrum: The art of doing twice the work in half the time" by Jeff Sutherland. Another fantastic book to look into is "Essential Scrum: A practical guide to the most popular agile process" by Kenneth S. Rubin. Both of those resources will tell you everything you need to know regarding the mechanics and philosophy of scrum.

FAQs

Do I need to be technical or know how to code in order to work as a scrum master?

There are multiple points that need to be made regarding this question. Scrum Masters are found in all type of industries, from event planning, to manufacturing, to telecommunications and of course software development. Some industries will not be as technical as others, but that should not discourage you from working in a technical field. Almost any business you work in will have a learning curve. If you do get a job in a very technical field, my advice is to ask a lot of questions and always carry a notebook with you to write down words and phrases that may be unfamiliar to you. Research and learn up on the new topics and you will slowly start piecing it all together. It is impossible to know everything, but it is completely reasonable to understand the 30,000 foot view of the industry you work in. I have never had more than a few programming courses myself, but I have made it a goal to ask a lot of questions to my team members and learn as much as possible about software development from various websites such as www.lynda.com, https://www.safaribooksonline.com/, and http://www.udemy.com.

What is the biggest challenge of being a Scrum Master?

Although there isn't necessarily one obstacle that will be the same for everyone, one of the biggest challenges is dealing with tough individuals. There will always be someone that you work with that does not want to be a team player or prefers isolating himself from everyone else. You may have multiple competing personalities that are always fighting for attention. As a Scrum Master, you have to find ways to work with each and every one of these types of people and encourage them to improve their teamwork and recognize the bigger picture. You may need to set up private discussions with these individuals and possibly bring in reinforcement such as a trusted advisor or manager that could help everyone find a resolution.

Do I need to follow every single recommendation in the scrum framework to be an effective Scrum Master?

No! The scrum framework was designed to give your teams the best chance to produce an MVP on a reoccurring basis. Going through a typical sprint should not feel mechanical, but rather like an opportunity to constantly create greatness. Flexibility is crucial when you are developing agile teams. For example, if for some reason, we can't hold a stand-up meeting, then we either post our statuses or we save it for the next day. Ultimately, you need to see what works best with your team and continue refining your methods and techniques. Continuous improvement is a motto that all agile teams should constantly strive for.

Final Thoughts

I hope you have enjoyed learning a little bit more about what a Scrum Master does on a daily basis. My typical week may be completely different from Scrum Masters in other industries, but many of the same principles and meetings will still be relevant. One of the coolest parts of my job is that no two days of work are ever the same. I have the opportunity to learn something new everyday and try experiments with my teams to find ways to constantly improve. If you are thinking about becoming a Scrum Master, it is crucial that you enjoy working with people of all backgrounds and skill levels and that you want to be a servant leader to others. I wish you the best on your agile journey and hope that this resource has given you clarity and the motivation to pursue a role as a Scrum Master.

-C.G. Williams

Made in the USA
San Bernardino, CA
01 May 2019